Process Development Symposium 2020

Online
1-3 June 2020

ISBN: 978-1-7138-1819-9

Printed from e-media with permission by:

Curran Associates, Inc.
57 Morehouse Lane
Red Hook, NY 12571

Some format issues inherent in the e-media version may also appear in this print version.

AIChE
120 Wall Street, FL 23
New York, NY 10005-4020

Phone: (800) 242-4363
Fax: (203) 775-5177

www.aiche.org

Additional copies of this publication are available from:

Curran Associates, Inc.
57 Morehouse Lane
Red Hook, NY 12571 USA
Phone: 845-758-0400
Fax: 845-758-2633
Email: curran@proceedings.com
Web: www.proceedings.com

TABLE OF CONTENTS

PROCESS DEVELOPMENT FOR A CIRCULAR ECONOMY

TECHNOLOGY TRANSFER KN A GLOBAL ENVIRONMENT

PANEL SESSION

MODELING, COMPUTATION, AND DATA ANALYTICS IN PROCESS DEVELOPMENT

EMERGING TECHNOLOGIES

CASE STUDIES

Sustainability and Corporate Responsibility in Process Development

Conference: Process Development Symposium

Year: 2020

Proceeding: 2020 Process Development Symposium

Group: General Submissions

Session: Opening Keynote Address

Time: Monday, June 1, 2020 - 9:10am-10:00am

Authors:

Raiford, W., The Chemours Company

Process development activities in the chemical industry continue to evolve, with extended focus on sustainability and how it shapes innovation, from discovery of new molecules and materials, though development and design of manufacturing processes, to commercialization of new offerings. Even the recent COVID-19 pandemic has informed how teams collaborate and manage complex development programs in a more virtual world. The Chemours Company established recently its 2030 Corporate Responsibility Commitment, a set of ambitious goals designed to deliver both social and environmental impact over the next decade. Within this framework, examples will highlight the critical role that sustainability is having on the innovation process, with an aim to motivate the 2020 PDS participants to focus their efforts to develop more sustainable solutions.

Modeling Heat Transfer for TNT Soil Remediation

Conference: Process Development Symposium

Year: 2020

Proceeding: 2020 Process Development Symposium

Group: General Submissions

Session: Challenges in Designing Sustainable Processes

Time: Monday, June 1, 2020 - 12:05pm-12:30pm

Authors:

Geiger, S., Chemours

Between 1905 and 1975, the DuPont Powder Company operated the largest explosives manufacturing plant in the world, producing over a half a billion pounds of explosives. The plant site, named Barksdale Works, was remotely located near Ashland, Wisconsin. Forty-five years after the site was closed and dismantled, there remains areas of soil that are contaminated with hazardous TNT, requiring effective remediation so that the land can be put to improved use. This presentation will detail an approach to large-scale soil TNT remediation that is an attractive and viable alternative to incineration. Considerations and challenges for a potential design will be discussed, with a specific focus on steady state and unsteady state heat transfer modeling.

Sustainability as an Exciting Innovation Challenge

Conference: Process Development Symposium

Year: 2020

Proceeding: 2020 Process Development Symposium

Group: General Submissions

Session: Challenges in Designing Sustainable Processes

Time: Monday, June 1, 2020 - 11:00am-11:25am

Authors:

Ressler, R., Corteva Agriscience

The world needs chemical engineers now as much as any time in history. We know that industrial processes have an impact on the environment, and, as the world's population continues to grow, we face greater challenges to sustain a healthy planet. Now, in the 21st century, we are equipped with technological advances in instrumentation, computing power and data analytics that are fueling great leaps in designing more sustainable chemical processes. Coupled with ever-advancing abilities to collaborate around the world and across disciplines, we are poised to make progress even more rapidly. This combination of the urgent need to solve the problems facing humanity, along with the encouraging prospect that we have the tools at our disposal to do just that, makes innovation for more sustainable chemical processes an exciting challenge. This talk will invite each of us to consider our respective industries and reimagine the future that we will create. In addition, the talk will highlight that by seeking a greater understanding of the entire value chain, we will be inspired to innovate and to find more sustainable solutions even faster. We have the great responsibility as chemical engineers to forge a new path for environmental sustainability. Pursuing these exciting solutions is a clear opportunity for innovation.

The key role of digital design in designing sustainable processes

Conference: Process Development Symposium

Year: 2020

Proceeding: 2020 Process Development Symposium

Group: General Submissions

Session: Challenges in Designing Sustainable Processes

Time: Monday, June 1, 2020 - 11:50am-12:05pm

Authors:

Leyland, S., Process Systems Enterprise Inc

Creating sustainable processes inevitably involves trade-offs that reconcile production, safety, profitability and sustainability objectives to minimize plant emissions and energy, water and other resources usage while achieving a range of production KPIs.

A digital design approach utilizes high-fidelity, predictive process models validated against experimental or pilot plant data to explore the process design space rapidly and effectively so that such decisions can be taken based on accurate quantification of the options. Key technologies are large scale Mixed Integer Nonlinear Programming (MINLP) optimization and Global System Analysis (GSA). MINLP optimization enables determination of the optimal value of multiple decision variables simultaneously while ensuring that equipment, process constraints, and safety constraints are observed. GSA allows rigorous quantification of the risk associated with the resulting optimal process design.

This presentation describes the digital design approach currently being applied to the design of processes from large-scale chemical plants to fuel cells to foodstuff production, briefly illustrated with real-world case studies.

The Life Cycle Assessment of Transforming Captured CO2 Into Chemical Products Using Electrochemical Catalyst Conversion Technology

Conference: Process Development Symposium

Year: 2020

Proceeding: 2020 Process Development Symposium

Group: General Submissions

Session: Challenges in Designing Sustainable Processes

Time: Monday, June 1, 2020 - 11:25am-11:50am

Authors:

Chou, J., Keylogic

The world needs chemical engineers now as much as any time in history. We know that industrial processes have an impact on the environment, and, as the world's population continues to grow, we face greater challenges to sustain a healthy planet. Now, in the 21st century, we are equipped with technological advances in instrumentation, computing power and data analytics that are fueling great leaps in designing more sustainable chemical processes. Coupled with ever-advancing abilities to collaborate around the world and across disciplines, we are poised to make progress even more rapidly. This combination of the urgent need to solve the problems facing humanity, along with the encouraging prospect that we have the tools at our disposal to do just that, makes innovation for more sustainable chemical processes an exciting challenge. This talk will invite each of us to consider our respective industries and reimagine the future that we will create. In addition, the talk will highlight that by seeking a greater understanding of the entire value chain, we will be inspired to innovate and to find more sustainable solutions even faster. We have the great responsibility as chemical engineers to forge a new path for environmental sustainability. Pursuing these exciting solutions is a clear opportunity for innovation.

Designing Robust & Flexible Processes: Perspective on Balancing Technology, Operational, & Economic Tradeoffs

Conference: Process Development Symposium

Year: 2020

Proceeding: 2020 Process Development Symposium

Group: General Submissions

Session: Designing Robust and Flexible Processes

Time: Monday, June 1, 2020 - 11:00am-11:25am

Authors:

Tieri, S., DuPont

Between 1905 and 1975, the DuPont Powder Company operated the largest explosives manufacturing plant in the world, producing over a half a billion pounds of explosives. The plant site, named Barksdale Works, was remotely located near Ashland, Wisconsin. Forty-five years after the site was closed and dismantled, there remains areas of soil that are contaminated with hazardous TNT, requiring effective remediation so that the land can be put to improved use. This presentation will detail an approach to large-scale soil TNT remediation that is an attractive and viable alternative to incineration. Considerations and challenges for a potential design will be discussed, with a specific focus on steady state and unsteady state heat transfer modeling.

Increasing Process Robustness and Flexibility with Integrated Continuous Manufacturing

Conference: Process Development Symposium

Year: 2020

Proceeding: 2020 Process Development Symposium

Group: General Submissions

Session: Designing Robust and Flexible Processes

Time: Monday, June 1, 2020 - 11:25am-11:50am

Authors:

Takizawa, B., CONTINUUS Pharmaceuticals, Inc.

For decades, pharmaceutical manufacturing did not evolve as other industries did. Process technologies largely remained the same, and innovation, while almost unbridled in drug discovery, was subordinated to established processes with minimal disruptive risks. But the risk of stagnation caught up with the industry, as outdated production systems failed quality requirements and revealed gross lead-time and cost inefficiencies. Fortunately, many drug manufacturers are now pivoting, some exploring how continuous manufacturing can lead to a paradigm more closely aligned with the expectations of regulators and consumers. The former have supported this movement, providing direction through draft guidance and regulations. One major advantage associated with this more advanced manufacturing method is the increased robustness it offers. For example, with continuous manufacturing, real-time monitoring of critical process parameters allows for immediate correction of process deviations, enabling a state of control. Conversely, with the current batch system, manufacturers rely on off-line tests that indicate whether a batch of material, usually large in quantity, meets specifications. These results are usually only available after the fact. Flexibility is another important consideration. Continuous unit operations are flexible and modular, and can be easily and quickly reconfigured to produce different drug products. Furthermore, their dial-up and dial-down capability enable production to adjust instantaneously to changes in demand. In this way, patient access can be improved, while maintaining a lean supply chain. The speaker will discuss these aspects of continuous manufacturing, along with other factors, focusing on Integrated Continuous Manufacturing (ICM), a strategy that integrates both upstream drug substance and downstream drug product manufacturing components into a single seamless process. Implementation challenges and potential solutions will also be explored.

Robustness and Flexibility in Pilot Plant Process Development Work

Conference: Process Development Symposium

Year: 2020

Proceeding: 2020 Process Development Symposium

Group: General Submissions

Session: Designing Robust and Flexible Processes

Time: Monday, June 1, 2020 - 11:50am-12:05pm

Authors:

Flores III, E., Southwest Research Institute

SwRI is fortunate to be part of chemical and petroleum process development programs that start from simple concept testing in the laboratory to operating commercial scale demonstration units. At SwRI we test and evaluate a broad range of technologies, which span the range from traditional refining and chemical unit operations to the newest methods for consuming plastics and carbon dioxide as feedstocks. For example, subjects have included: thin film polymer commercial demonstration, natural gas to olefins, and carbon capture and conversion pilot plants. With over 40 years of experience in pilot plant process development, we have had the opportunity to learn what is successful for our clients when it comes to designing robust and flexible processes. Many clients need more than technoeconomic evaluations (or "paper studies"), but rather require actual process data at a scale that is reliable for industrial uses such as pilot plant data. Pilot plants are unique in that engineers typically have to become "experts" very quickly in the broad range of chemical engineering work to be conducted in the test system. Pilot plants are designed for lower-cost testing or proof-of-concept evaluation, so resources are typically not allocated to provide commercially applicable designs for each unit operation or piece of equipment as might be used in commercial units. Process development is a feedback process, and pilot plants require innovative and flexible designs to accommodate ranges of process conditions that may not be applicable to commercial scale processes. Given the feedback part of process development, the robustness and flexibility are only as good as your last set of data. SwRI has compiled our learnings and developed a procedure for process development called the SwIFT (SwRI Interdependent Flexible Time-driven) process for information development. SwIFT process development is fundamentally about derisking a process and, if necessary, making as many mistakes as you can early on, at the smallest believable scale. In a way, pilot-scale tests are insurance against troubleshooting on an already-built design that is resisting a successful startup. While there is no replacement for experience and existing infrastructure, the SwIFT process uses available experience (with constant learning) while pressing forward with process development. SwRI will discuss some of the "good and bad" aspects when it comes to designing robust and flexible processes at pilot scale.

Keep Your Cool: Rapid Development to Address an Upset Scenario During a 1st API Delivery

Anne E. Mohan

Session 2: Designing Robust and Flexible Processes

Jun 1, 2020

AIChE — The Global Home of Chemical Engineers | pds PROCESS DEVELOPMENT SYMPOSIUM | MERCK INVENTING FOR LIFE

Anne E. Mohan – about the speaker

- Born and raised in Brooklyn, NY
- Manhattan College graduate, B.S. 2009, M.S. 2010
- Merck & Co., Inc. since August 2010, Rahway, N.J.
  - All tenure in Chemical Engineering R&D
    - Department works on small molecule process development and scale-up for the Active Pharmaceutical Ingredient (API)
    - Most recent ~2.5 years in Crystallization Lab specialty group
  - Recently moved to Biologics Process R&D to work on mAbs crystallizations

CERD — Bridging Research & Reality

- Develop robust, scalable crystallization processes that produce quality API for clinical trials and ultimately commercial launch
- Job function is a combination of:
  - Lab-scale process development (5 – 10000 mL)
  - Pilot plant campaigns to supply API for clinical trials
  - Process characterization to enable validation and launch

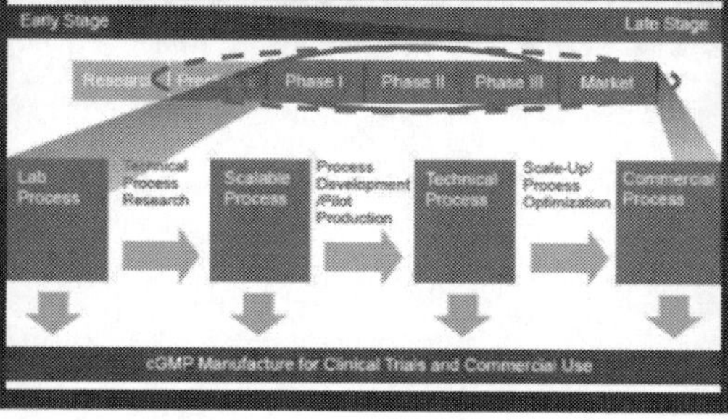

AIChE | pds | MERCK INVENTING FOR LIFE

Pharma Project Background

- Challenges for the API crystallization:
  - Propensity of API crystal to form solvates
  - Particle size control requirements unknown
  - Variability in starting material purity
- Execute targeted, rapid process development to support:
  - Physical characterization of the desired crystal form
  - 1st GMP delivery of API which will supply clinical trials
  - Collaboration between drug substance and drug product to understand PSD requirements (if any)

Drug substance

Drug product

AIChE pds MERCK INVENTING FOR LIFE

Kilo Lab Demo, 1st API Delivery: Pure Step Isolation

- Process Overview:
  - Dissolve starting material in IPAc
  - Charge antisolvent to generate supersaturation
  - Charge seed and age seed bed
  - Charge antisolvent
  - Filter, wash, and dry

End seed bed age

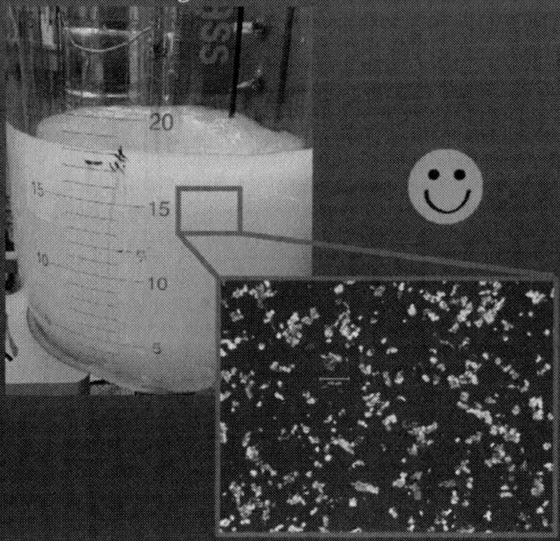

AIChE pds MERCK INVENTING FOR LIFE

Kilo Lab Demo, 1st API Delivery: Pure Step Isolation

- Process Overview:
  - Dissolve starting material in IPAc
  - Charge antisolvent to generate supersaturation
  - Charge seed and age seed bed
  - Charge antisolvent
  - Filter, wash, and dry

90 minutes into antisolvent charge

AIChE | pds | MERCK INVENTING FOR LIFE

Investigation of Unexpected Morphology

<u>90-min Slurry Analysis:</u>

- XRPD does not match any known forms
- TGA-IR weight loss of initial sample doesn't exactly match theoretical values for mono or hemi solvate, <u>but</u> NMR shows only IPAc, so must be some version of an IPAc solvate!

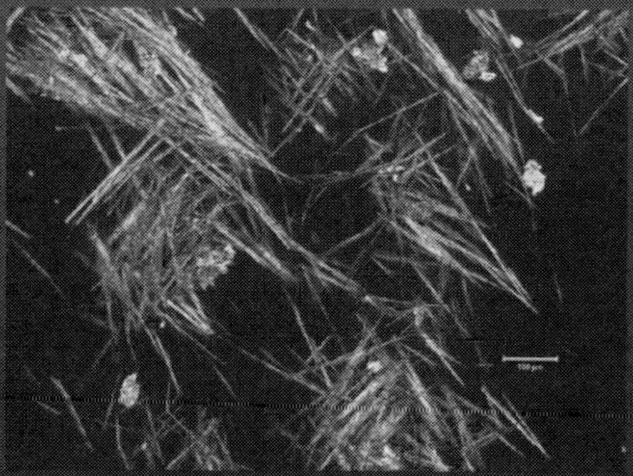

6

AIChE | pds | MERCK INVENTING FOR LIFE

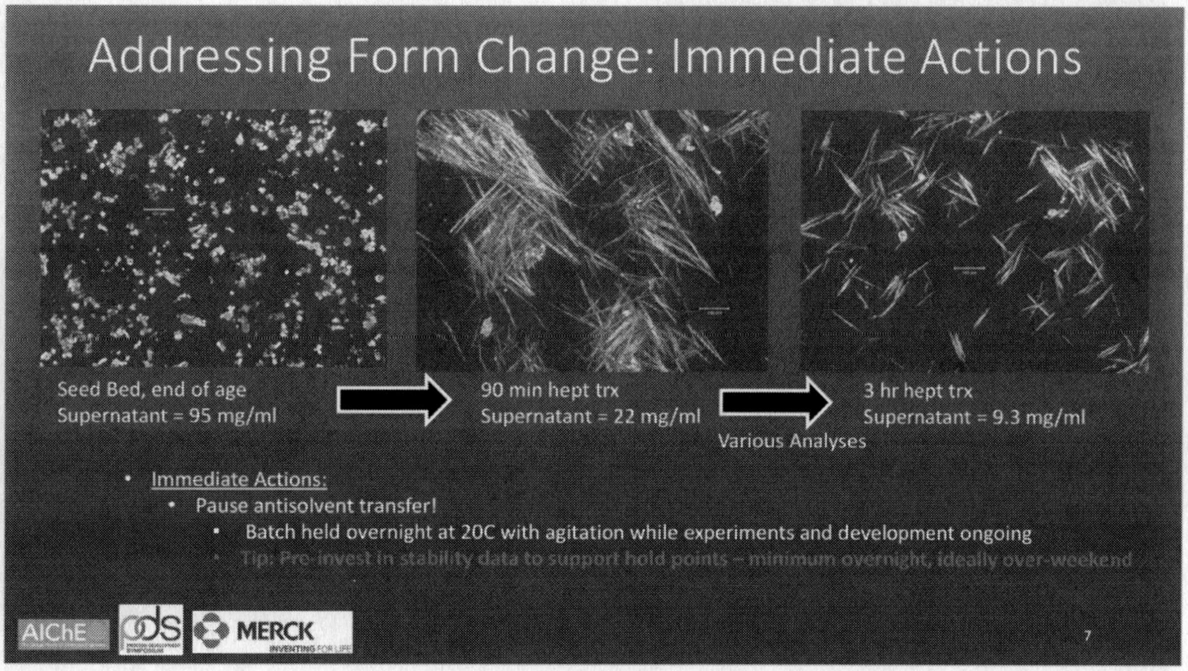
Addressing Form Change: Immediate Actions
Seed Bed, end of age
Supernatant = 95 mg/ml
90 min hept trx
Supernatant = 22 mg/ml
3 hr hept trx
Supernatant = 9.3 mg/ml
Various Analyses
Immediate Actions:
Pause antisolvent transfer!
Batch held overnight at 20C with agitation while experiments and development ongoing
Tip: Pre-invest in stability data to support hold points – minimum overnight, ideally over-weekend
AIChE
pds
MERCK
INVENTING FOR LIFE
7

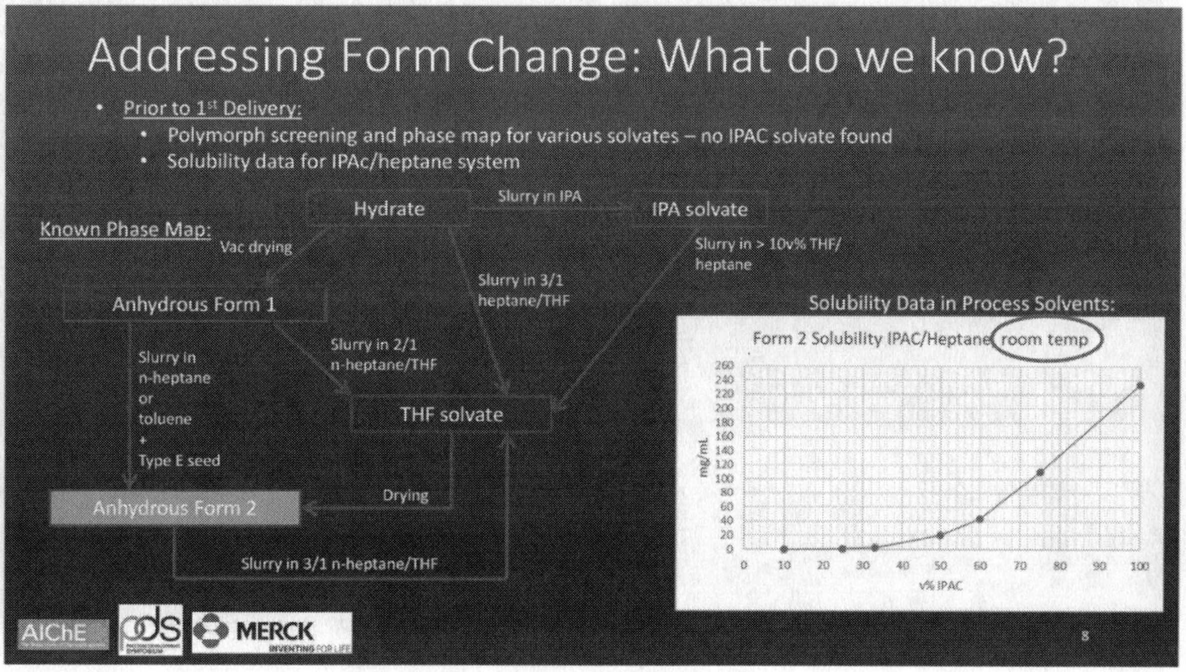
Addressing Form Change: What do we know?
Prior to 1st Delivery:
Polymorph screening and phase map for various solvates – no IPAC solvate found
Solubility data for IPAc/heptane system
Known Phase Map:
Hydrate
Slurry in IPA
IPA solvate
Vac drying
Slurry in > 10v% THF/heptane
Slurry in 3/1 heptane/THF
Anhydrous Form 1
Slurry in 2/1 n-heptane/THF
Slurry in n-heptane or toluene + Type E seed
THF solvate
Drying
Anhydrous Form 2
Slurry in 3/1 n-heptane/THF
Solubility Data in Process Solvents:
Form 2 Solubility IPAC/Heptane room temp
mg/mL
v% IPAC
AIChE
pds
MERCK
INVENTING FOR LIFE
8

Rapid, Parallel Process Development

	Yielded Form 2 (desired)	Yielded IPAc Solvate (Undesired)
To Determine the Path Forward	• Prep lab slurry heated to 55C, held overnight • Prep lab slurry heated to 40C, 55C, & 80C; Form 2 seed charged, aged overnight • Prep lab slurry heated to 70C; Form 2 seed charged; Cool to 20C over 12 hours	• Prep lab slurry filtered/washed, then dried overnight at 40C, 80C, & 100C
To Understand the Phase Map	• Seed bed prep at 30C, charge n-heptane at 30C	• Seed bed prep at 10C, charge n-heptane at 10C

RAHWAY SITE STEAM SHUTDOWN NOTIFICATION

AIChE | pds | MERCK INVENTING FOR LIFE

9

Rapid, Parallel Process Development

	Yielded Form 2 (desired)	Yielded IPAc Solvate (Undesired)
To Determine the Path Forward	• Prep lab slurry heated to 55C, held overnight • Prep lab slurry heated to 40C, 55C, & 80C; Form 2 seed charged, aged overnight • Prep lab slurry heated to 70C; Form 2 seed charged; Cool to 20C over 12 hours ★	• Prep lab slurry filtered/washed, then dried overnight at 40C, 80C, & 100C
To Understand the Phase Map	• Seed bed prep at 30C, charge n-heptane at 30C ★	• Seed bed prep at 10C, charge n-heptane at 10C

★ <u>Competitive slurry experiments:</u>

- Charged IPAC solvate to both slurries, age overnight
  - 30C slurry → Form 2 ☺
  - 20C slurry → Form 2 + solvate ☹

AIChE | pds | MERCK INVENTING FOR LIFE

10

Rapid, Parallel Process Development

	Yielded Form 2 (desired)	Yielded IPAc Solvate (Undesired)
To Determine the Path Forward	• Prep lab slurry heated to 55C, held overnight • Prep lab slurry heated to 40C, 55C, & 80C; Form 2 seed charged, aged overnight ★ • Prep lab slurry heated to 70C; Form 2 seed charged; Cool to 20C over 12 hours	• Prep lab slurry filtered/washed, then dried overnight at 40C, 80C, & 100C
To Understand the Phase Map	• Seed bed prep at 30C, charge n-heptane at 30C ★	• Seed bed prep at 10C, charge n-heptane at 10C

★ Competitive slurry experiments:

- Charged IPAC solvate to both slurries, age overnight
  - 30C slurry → Form 2
  - 20C slurry → Form 2 + solvate

- Conclusions:
  - Temperature dependency of phase map for IPAc solvate and Form 2

AIChE | pds | MERCK INVENTING FOR LIFE

11

Using Process Analytical Technology

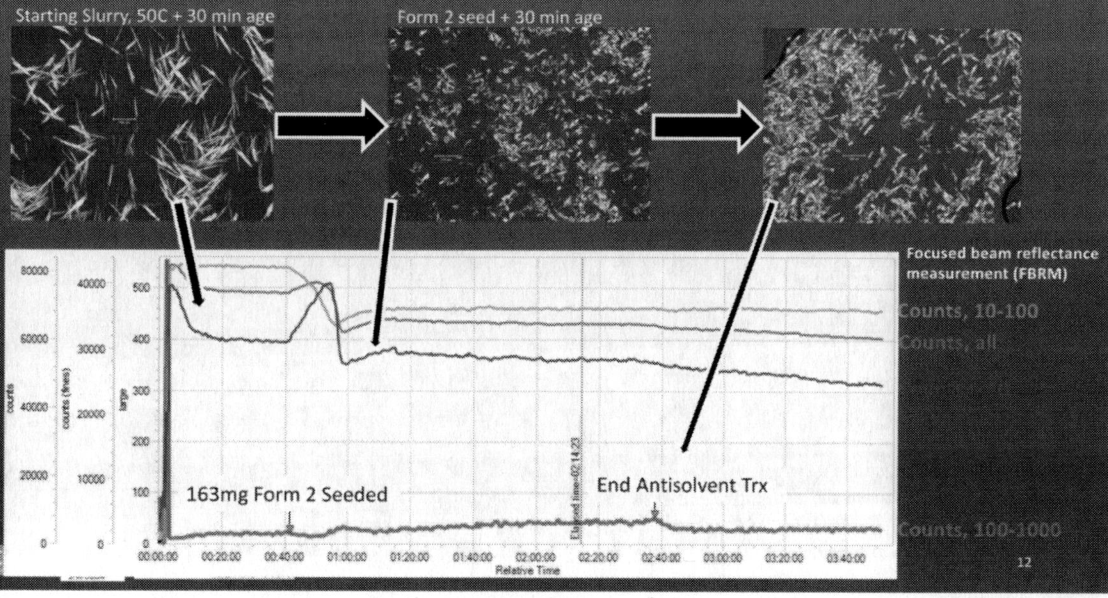

12

Prep Lab Batch Path Forward and Post-Processing Knowledge

- Based on all these results, procedure for prep lab:
  - Higher temperature processing + Form 2 seeding
  - Vessel to vessel transfer for heptane addition (jacketed ability)
- 1.5 kg isolated dry cake – XRPD conforms to Form 2!

- Post 1st Delivery Data:
  - Phase map for IPAC solvate vs Form 2 – confirmed temperature dependency!
  - Guidance for 2nd delivery, which was ~1 week behind 1st delivery

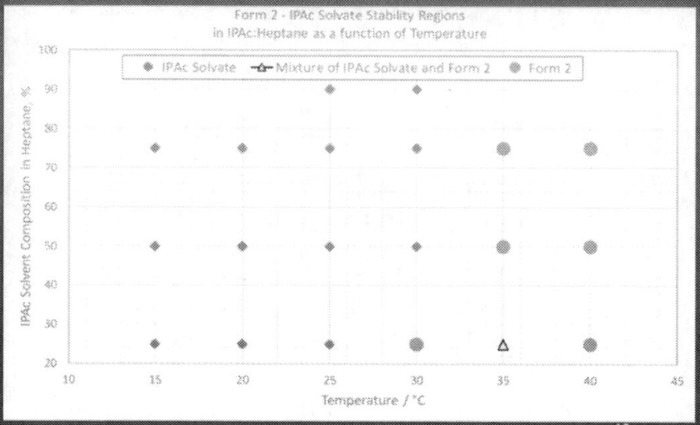

AIChE | pds | MERCK INVENTING FOR LIFE

13

Learnings, Conclusions, and Recommendations

- Scale-Up Guidelines:
  - Due diligence for existing data to understand current state
  - Pre-invest in stability data!
  - Review (and re-review) physical set-up to highlight any differences between lab and scale-up
  - Think about physical environment of scale-up!
    - "Ambient" in one area might not = "ambient" in another!

- Upset Scenario Guidelines
  - Cross-functional support to execute orthogonal analyses of upset scenario
  - Lab development objectives balanced between mechanistic understanding and operational directives

- Post Processing Guidelines:
  - After-action reviews (varying degrees of formality)

AIChE | pds | MERCK INVENTING FOR LIFE

Acknowledgements

Process Chemistry
Caleb Hethcox
Tetsuji Itoh
Jungchul Kim
Diane Le

Materials Characterization
Laura Artino

Analytical R&D
Holst Halsey
Brittany Kassim
Matthew Gunsch

External Capabilities
Yonggang Chen
Lushi Tan

Chem Eng R&D
Mark Weisel
Surendra Gadamsetty
Lisa Frey
Vito Rapisardi
Noberto Rivera

Thank you for your attention! Questions?

Science cannot solve the ultimate mystery of nature. And that is because, in the last analysis, we ourselves are a part of the mystery that we are trying to solve.

(Max Planck)

AIChE pds MERCK INVENTING FOR LIFE

Enabling Technologies Consortium: A new paradigm for pre-competitive collaboration on technology

Conference: Process Development Symposium

Year: 2020

Proceeding: 2020 Process Development Symposium

Group: General Submissions

Session: Poster Session

Time: Monday, June 1, 2020 - 5:06pm-5:24pm

Authors:

Ferretti, A. C., Celgene

The Enabling Technologies Consortium (www.etconsortium.org), as a forum for pharmaceutical companies to discuss ideas, share information, and collaborate on the development of new enabling technologies used in pharmaceutical drug substance and product development, has now been operating since 2017. ETC gives member companies the ability to partner with external collaborators to further the mission of developing pre-competitive enabling technologies for pharmaceutical development and manufacturing. Pre-competitive projects, defined here as collaborations between multiple pharmaceutical companies and third parties, are in progress in the areas of solubility, crystallization, modeling, analytical instrumentation, and high throughput automation. The goal is to produce an efficiency-enhancing advancement or refinement that will ultimately be made broadly available to the public, either through publication, commercialization of a new product, or other means. ETC has been able to streamline these partnerships by eliminating the need for parallel agreements between each member company and the external party, providing a governance framework for the advancement of collaborative projects, and allowing for companies interested in enabling technologies to invest equally in their development. ETC members are also able to provide feedback on newly developed technologies upon request from vendors. This poster showcases the how ETC works and the impact it can have in the area of process development for Drug Substance and Product development.

Learnings from a Career in Process Development

Conference: Process Development Symposium

Year: 2020

Proceeding: 2020 Process Development Symposium

Group: General Submissions

Session: Poster Session

Time: Monday, June 1, 2020 - 5:30pm-5:45pm

Authors:

Sisler, G., DuPont

Gregg is a Process Development Engineer with experience at the full range of scale – from lab through commercial - including startups. He was trained as a Chemical Engineer at the University of Minnesota and the University of California, Berkeley. In his career with DuPont and INVISTA, Gregg helped commercialize new technologies in THF and poly-THF (Terathane®) in LYCRA® intermediates, was Technology Guardian in Aramid Intermediates for Nomex® and Kevlar®, scaled-up fire-fighting foams and surface-protection agents in Fluoroproducts, and developed and commercialized OLEDs and Display Materials offerings in Electronics and Communications. Gregg gained experience in hydrogenation, oxidation, catalyst development, chemical synthesis and polymerization processes, and in handling hazardous materials, including butane, ethylene oxide and phosgene. He is expert in basic-data development and process safety and has learned to interface with management, technical experts, colleagues, and operators . This poster will review some of the successes and challenges of his endeavors in process development and scale-up and attempt to characterize why things went well in some programs and why they didn't in others.

Solvent Separation From Silicone Resin: A Case Study in Process Route Selection

Conference: Process Development Symposium

Year: 2020

Proceeding: 2020 Process Development Symposium

Group: General Submissions

Session: Poster Session

Time: Monday, June 1, 2020 - 4:45pm-5:00pm

Authors:

Taylor, L., The Dow Chemical Company

Amid business pressure for rapid commercialization, route selection is easily overlooked to focus ever limited resources on fast implementation. However, this tendency can lead to expensive, inefficient, and unreliable processes. Even brief comparison of multiple process routes can yield a more efficient, sustainable, and robust process for decades of successful manufacturing.

This poster examines a case study in route selection around a silicone resin-solvent separation. The highly viscous nature of the resinous material challenged traditional separation techniques. First, the team proposed pre-dilution with low viscosity silicone fluids to enable consideration of multiple separation techniques. Next, an alternative separation technique was recommended to improve manufacturing capability, conversion cost, and quality consistency. Finally, a mid-process, partial dilution was planned to create a leverageable intermediate, reduce contamination risk, and maximize bulk packaging flexibility.

Early-stage consideration of safety, sustainability, feasibility, cost, supply chain, timing, and risks ultimately led to implementation of a robust, and profitable process. Furthermore, transitioning from the presumed process route to an alternate route reduced manufacturing risk, process waste, and even time to launch by avoiding failed scale-up attempts with an incapable process. Overall, this process development example demonstrates the importance of balancing adequate process route consideration with swift implementation for the greatest business impact.

Process Troubleshooting and Process Technology Development via Reaction Calorimetry

Conference: Process Development Symposium

Year: 2020

Proceeding: 2020 Process Development Symposium

Group: General Submissions

Session: Poster Session

Time: Monday, June 1, 2020 - 4:45pm-5:00pm

Authors:

Jeske, S., The Dow Chemical Company
Deeth, B., The Dow Chemical Company

Reaction calorimetry is a standard lab measurement technique to understand the energy released or absorbed due to chemical reactions. Energy transferred from a chemical reaction, in the form of heat, is directly proportional to the rate of reaction. Consequently, calorimetry measurements can be used to study reaction kinetics and assess the impact of process variations.

Production of some silicone fluids involve highly exothermic chemistries and these chemistries can be very sensitive to changes in raw material impurities. When production of an established silicone fluid process began to exhibit out of control product quality signals, a quick assessment of the production process conditions revealed that the reaction temperature profile had changed. To accomplish a process evaluation, the production process was scaled-down to the lab scale and reaction calorimetry was employed to study the reaction kinetics. Analysis of the resulting experimental data was used to identify a production change in one of the raw materials as the root cause.

By understanding and monitoring kinetics through reaction calorimetry, a process technology to mitigate the impact of a trace impurity was developed and scaled up to the production process. This poster outlines the use of reaction calorimetry to troubleshoot a production upset and its use to develop a process technology that improved process robustness due to raw material variation.

Carbon Recycling with LanzaTech's Gas Fermentation Process

Conference: Process Development Symposium

Year: 2020

Proceeding: 2020 Process Development Symposium

Group: General Submissions

Session: Process Development for a Circular Economy

Time: Tuesday, June 2, 2020 - 9:25am-9:50am

Authors:

Gao, A., Lanzatech

LanzaTech's innovative commercial scale gas fermentation platform can capture and utilize CO_2, H_2 and CO for the production of fuels and chemicals. The ability of the fermentation platform to utilize gas mixtures of varying compositions enables the conversion of a diverse range of feedstocks, including waste gases from industrial sources (e.g. refineries and steel mills) and syngas generated from biomass or municipal solid waste.

Developing a robust process capable of handling the fluctuations present in waste gas streams requires piloting on live gas streams at the production site. LanzaTech has operated for greater than 100,000 hours on live gas at various pilot and demonstration units, resulting in commercialization of the gas fermentation process in 2018 with the startup of the Shougang LanzaTech plant in Caofeidian, China. Two other plants operating on industrial off-gases are under construction in Belgium (ArcelorMittal Gent, steel off-gas) and India (Indian Oil Company, refinery off-gas). Looking beyond industrial off-gas, gasification of municipal solid waste and conversion of syngas in gas fermentation offers an alternative pathway to recycling, where wastes which are not amenable to mechanical recycling can broken down and reformed into virgin materials.

Circles, lines and bars: bringing life cycle thinking to evaluating circular economy concepts

Conference: Process Development Symposium

Year: 2020

Proceeding: 2020 Process Development Symposium

Group: General Submissions

Session: Process Development for a Circular Economy

Time: Tuesday, June 2, 2020 - 9:00am-9:25am

Authors:

Helling, R. K., The Dow Chemical Company

Many of the materials and chemical in society are used in linear fashion – they are made, used and disposed of. There is growing interest in steering society towards more circular value chains: make, use, and then reuse, repurpose or recycle. Using materials in a more circular way can mimic natural processes and may be more sustainable than more linear ways. Emerging metrics for and evaluations of "circularity" of a process, product or company often focus exclusively on material flows. Sustainability is a broader concept which includes not only consideration of material flows, but also energy use, emissions and the impact of these. Life cycle assessment (LCA) is a methodology that is well suited for creating a comprehensive evaluation of the potential environmental impacts of new "circular" products or processes. This presentation will use LCA to evaluate a hypothetical process (or processes) for recycling waste plastics back into "good as virgin" plastics. The example will show the utility of LCA as a tool an approach to clearly define under which conditions a circular process is indeed more sustainable and when it is not. This insight can be used to define the critical conditions beyond the process technology (material supply, power sources, geography, other waste management) that can define success.

Process Design and Demonstration of the Plasco GPRSTM: A zero-waste omnivorous conversion process for the circular economy

Conference: Process Development Symposium

Year: 2020

Proceeding: 2020 Process Development Symposium

Group: General Submissions

Session: Process Development for a Circular Economy

Time: Tuesday, June 2, 2020 - 9:50am-10:15am

Authors:

Bacon, M., Plasco Conversion Technologies

The Plasco Gasification and Plasma Refining System (GPRSTM) was developed and demonstrated at commercial scale for over a decade in Ottawa, Ontario for the purpose of maximizing the conversion of energetic waste to valuable products. It converts minimally prepared solid waste to an ultraclean syngas fuel, hot air, non-leachable slag, and recovers water from the waste for sewer or surface discharge.

The process is omnivorous. Incoming wastes can include Municipal Solid Waste (MSW), Refuse-Derived Fuel (RDF), residual biomass from forestry and agriculture, and hazardous wastes. The GPRS is configured in compact 200 tpd process trains. It produces no emissions to atmosphere during conversion. The performance of the unit is guaranteed by a technology performance insurance from New Energy Risk, a division of AXA.

The main markets include making a precursor for hydrogen, liquid fuels and plastics, powering IC engines and turbines, replacing natural gas in industrial applications and treating hazardous wastes.

PlascoSyngasTM has a lower heating value (LHV) that can be tailored across a wide range to suit the end use. For each application, the produced syngas is controlled to consistent and predictable composition and flowrate. Tars are reduced by 99.95%, pH is balanced, sulfur and contaminants are removed to low levels, temperature and relative humidity are tightly controlled, and pressure is regulated. Dioxins and furans are below the level of quantification.

Cold gas efficiency is high, and parasitic power usage low. There is no residue sent to landfill.

The paper will explain how tars are destroyed, particulate is completely captured and converted to vitrified slag of commercial value, and dioxin and furan formation is avoided. Examples of projects for making syngas as fuel and hydrogen precursor from biomass are described, as well as a proposal for completely diverting MSW from landfill while making H_2 at prices competitive with gasoline and diesel fuel.

A Virtual R&D Center for High-Energy Materials Testing and Analysis

Conference: Process Development Symposium

Year: 2020

Proceeding: 2020 Process Development Symposium

Group: General Submissions

Session: Technology Transfer in a Global Environment

Time: Tuesday, June 2, 2020 - 11:00am-11:25am

Authors:

Brandes, T., Chemours

Characterization of high-energy systems and materials is a vital part of a chemical company's responsible and sustainable execution of product development, operation, and change management. When Chemours was spun-off from DuPont, the lab facilities and some of the technical expertise remained with DuPont, leaving Chemours with the need to re-establish similar capabilities. The most expedition solution was to establish a virtual R&D center for high energy materials testing and analysis. This presentation will discuss the steps taken to survey and define testing needs, collaboration with 3rd-party labs to establish contractual access to testing facilities, and the creation of a formal workflow to interface the remotely-located program principal engineer with internal customers throughout Chemours. A case example is also provided for a what-if scenario identified in a PHA for an out-of-order addition of catalyst to a highly reactive system, how testing was designed and executed, and how the results were delivered back to the customer.

Importance of Culture in Process Careers

Conference: Process Development Symposium

Year: 2020

Proceeding: 2020 Process Development Symposium

Group: General Submissions

Session: Technology Transfer in a Global Environment

Time: Tuesday, June 2, 2020 - 11:50am-12:15pm

Authors:

Shepard, L., The Dow Chemical Company

Technology transfer, by definition, involves a change in location, scale, equipment details, and other facility-specific conditions. Many of these aspects can be addressed by documents commonly used in technology transfer such as process material and energy balances, operating protocols, equipment design requirements, site specific checklists. These documents, however, are often incomplete or misunderstood as a result of information gaps or unidentified assumptions. Additionally, language and cultural differences often create barriers to uncovering these information gaps and misunderstandings. Deployment of key project management practices, such as Risk Assessment, RACI, Integrated Project Planning, document sharing and regular communication cadences, can provide a framework within which to mitigate communication barriers. For example, thoughtful development of the project Risk Register can ensure that everyone's concerns are heard and documented, which then opens subsequent communication. Document sharing provides a time-flexible channel to accumulate written information in a way that makes limited interaction time more productive. Regular communication cadences provide a forum for timely airing of questions and issues, and reiteration of critical points as needed. In deploying all these techniques, it is important to keep in mind the value of communicating the same information or asking key questions in multiple formats to ensure that the correct and critical information comes to light.

Technology Transfer: Bridging the Communication Gaps

Conference: Process Development Symposium

Year: 2020

Proceeding: 2020 Process Development Symposium

Group: General Submissions

Session: Technology Transfer in a Global Environment

Time: Tuesday, June 2, 2020 - 11:25am-11:50am

Authors:

Threefoot, K., DuPont

Technology transfer, by definition, involves a change in location, scale, equipment details, and other facility-specific conditions. Many of these aspects can be addressed by documents commonly used in technology transfer such as process material and energy balances, operating protocols, equipment design requirements, site specific checklists. These documents, however, are often incomplete or misunderstood as a result of information gaps or unidentified assumptions. Additionally, language and cultural differences often create barriers to uncovering these information gaps and misunderstandings. Deployment of key project management practices, such as Risk Assessment, RACI, Integrated Project Planning, document sharing and regular communication cadences, can provide a framework within which to mitigate communication barriers. For example, thoughtful development of the project Risk Register can ensure that everyone's concerns are heard and documented, which then opens subsequent communication. Document sharing provides a time-flexible channel to accumulate written information in a way that makes limited interaction time more productive. Regular communication cadences provide a forum for timely airing of questions and issues, and reiteration of critical points as needed. In deploying all these techniques, it is important to keep in mind the value of communicating the same information or asking key questions in multiple formats to ensure that the correct and critical information comes to light.

Innovation & Sustainability in Process Development

Conference: Process Development Symposium

Year: 2020

Proceeding: 2020 Process Development Symposium

Group: General Submissions

Session: Panel Session

Time: Tuesday, June 2, 2020 - 2:00pm-3:30pm

Authors:

Sarafinas, A., The Dow Chemical Company

Fluidized beds (FB) and circulating fluidized beds (CFB) are widely used in industry. The major processes that utilize FBs and CFBs will be discussed. Each process has evolved over the years with its own design rules and culture. Understanding various processes and how these unit operations are implemented help us understand the best practices and general design guidelines.

This webinar will cover the following topics:

- Particle size, density and Geldart Group classification,
- Unifying Regime phase diagram indicating various regimes of fluidization,
- Fixed bed vs. fluidized beds vs. circulating fluidized beds,
- Catalytic processes,
- Noncatalytic processes,
- Physical processes,
- Regimes of fluidization in each of this process

Computational-Aided Pharmaceutical Process Development in the Era of a Global Pandemic

Conference: Process Development Symposium

Year: 2020

Proceeding: 2020 Process Development Symposium

Group: General Submissions

Session: Modeling, Computation, and Data Analytics in Process Development

Time: Wednesday, June 3, 2020 - 9:00am-9:25am

Authors:

Luciani, C., Eli Lilly & Co.

Synthetic molecules continue to play a major role in the development of treatments that benefit the lives of patients. However, the landscape of synthetic molecules has changed significantly over the last 30 years. From a rich small-molecule portfolio to a combination of small and medium size molecules, drug substance complexity, potency, and processing platforms have been evolving to adapt to the patient needs. Low volume/high potent/highly complex active ingredients define new design requirements. Practitioners responded to this trend by adopting advanced manufacturing approaches. This created the need for an interesting complementary that could speed up the discovery/development of novel drugs. Modeling and simulation started to gain traction.

This talk will describe the journey of drug substance process development through the lens of mechanistic models.

The advances made over the last few years are key for the development of new molecular entities and routine manufacturing of commercial medicines. This is showing to critical also to meet the need of ultra-rapid discovery, development, and manufacturing as a response to a biological threat such as COVID-19.

Population Balance Modeling to Predict Particle Size Distribution Upon Scale-up of a Combined Antisolvent and Cooling Crystallization

Conference: Process Development Symposium

Year: 2020

Proceeding: 2020 Process Development Symposium

Group: General Submissions

Session: Modeling, Computation, and Data Analytics in Process Development

Time: Wednesday, June 3, 2020 - 10:05am-10:30am

Authors:

Rosenbaum, T., Bristol-Myers Squibb

In many agitated tanks and fluid handling systems, yield and performance is governed by complex turbulent fluid mechanics. These turbulent motions, which can inform both bulk transport and reaction kinetics, occur over a large range of length and time-scales. From a modeling perspective, direct numerical simulation of this entire turbulence spectrum is required to obtain complete insights into system performance. Due to time and resource constraints, however, such detailed simulations have not historically been practical within most industrial settings.

In this work, we show how graphical processing units (GPUs) can make large eddy simulations (LES) direct numerical simulation (DNS) of industrial mixing systems practical and timely. Although GPUs have historically been used for image rendering, in recent years they have emerged and as powerful computational rivals to traditional CPUs. We will begin the presentation by introducing the concepts governing fluid modeling on GPUs. We then show how, given identical algorithms, a single scientific GPU can execute simulations two to three orders-of-magnitude faster than a single CPU. We then present various criteria for monitoring the convergence from large-eddy simulation (LES) to direct numerical simulation, within the context of the turbulent fluid motion. We then apply these criteria to study blending, energy dissipation, and reaction rates in a bioreactor.

Beyond improvements in physical fidelity, GPU-based computational approaches enable real-time, virtual reality rendering of bioreactors and fluid handling equipment. Thus, in addition to providing basic numerical data, visual output can be piped real-time to a virtual reality headset for immersive insights into the transport processes ongoing inside a bioreactor.

Using Digital Twins to Optimize Engineering Processes

Conference: Process Development Symposium

Year: 2020

Proceeding: 2020 Process Development Symposium

Group: General Submissions

Session: Modeling, Computation, and Data Analytics in Process Development

Time: Wednesday, June 3, 2020 - 9:50am-10:05am

Authors:

Thomas, J. A., M-Star Simulations

In many agitated tanks and fluid handling systems, yield and performance is governed by complex turbulent fluid mechanics. These turbulent motions, which can inform both bulk transport and reaction kinetics, occur over a large range of length and time-scales. From a modeling perspective, direct numerical simulation of this entire turbulence spectrum is required to obtain complete insights into system performance. Due to time and resource constraints, however, such detailed simulations have not historically been practical within most industrial settings.

In this work, we show how graphical processing units (GPUs) can make large eddy simulations (LES) direct numerical simulation (DNS) of industrial mixing systems practical and timely. Although GPUs have historically been used for image rendering, in recent years they have emerged and as powerful computational rivals to traditional CPUs. We will begin the presentation by introducing the concepts governing fluid modeling on GPUs. We then show how, given identical algorithms, a single scientific GPU can execute simulations two to three orders-of-magnitude faster than a single CPU. We then present various criteria for monitoring the convergence from large-eddy simulation (LES) to direct numerical simulation, within the context of the turbulent fluid motion. We then apply these criteria to study blending, energy dissipation, and reaction rates in a bioreactor.

Beyond improvements in physical fidelity, GPU-based computational approaches enable real-time, virtual reality rendering of bioreactors and fluid handling equipment. Thus, in addition to providing basic numerical data, visual output can be piped real-time to a virtual reality headset for immersive insights into the transport processes ongoing inside a bioreactor.

Virtual process development using modeling & simulation: some examples

Conference: Process Development Symposium

Year: 2020

Proceeding: 2020 Process Development Symposium

Group: General Submissions

Session: Modeling, Computation, and Data Analytics in Process Development

Time: Wednesday, June 3, 2020 - 9:25am-9:50am

Authors:

Longwell, G., Siemens

Synthetic molecules continue to play a major role in the development of treatments that benefit the lives of patients. However, the landscape of synthetic molecules has changed significantly over the last 30 years. From a rich small-molecule portfolio to a combination of small and medium size molecules, drug substance complexity, potency, and processing platforms have been evolving to adapt to the patient needs. Low volume/high potent/highly complex active ingredients define new design requirements. Practitioners responded to this trend by adopting advanced manufacturing approaches. This created the need for an interesting complementary that could speed up the discovery/development of novel drugs. Modeling and simulation started to gain traction.

This talk will describe the journey of drug substance process development through the lens of mechanistic models.

The advances made over the last few years are key for the development of new molecular entities and routine manufacturing of commercial medicines. This is showing to critical also to meet the need of ultra-rapid discovery, development, and manufacturing as a response to a biological threat such as COVID-19.

Chemical Looping Processes of Renewable Fuel and Chemical Synthesis

Conference: Process Development Symposium

Year: 2020

Proceeding: 2020 Process Development Symposium

Group: General Submissions

Session: Emerging Technologies

Time: Wednesday, June 3, 2020 - 11:00am-11:25am

Authors:

Tong, A., The Ohio State University

The Biomass-to-Syngas (BTS) chemical looping process is an efficient thermochemical gasification process for the conversion of biomass to syngas via partial oxidation. The BTS process combines an iron-titanium composite metal oxide (ITCMO) oxygen carrier and a co-current gas-solid packed moving bed reducer design. The thermodynamics of the ITCMO oxygen carrier favors the controlled partial oxidation of the biomass feedstock and, hence, the generation of high purity syngas with minimal full oxidation products. The gas-solid co-current moving bed reducer operation ensures a sufficient gas and solid residence time for the full conversion of biomass volatiles and char, resulting in a syngas composition that is close to the thermodynamic equilibrium. The addition of steam to the system can adjust the composition of the syngas product to a desired H_2:CO ratio for the downstream chemical and fuel synthesis. In addition to the BTS process, the chemical looping hydrogen generation (CLHG) process is effective method to converted renewable fuels, such as biogas, to high purity, high pressure renewable hydrogen. The CLHG process uses a counter current gas-solid contacting pattern in a packed moving bed reactor to convert biogas to CO_2 in on reactor (i.e. the reducer) and another counter current moving bed reactor to convert steam to H_2 (i.e. the oxidizer). The BTS and CLHG processes have been experimentally tested in a bench scale moving bed reducer and oxidizer. In the BTS process, the composition of syngas, including tar content, is quantified. In the CLHG, the composition of CO_2 from the reducer and H_2 from the oxidizer have been studied. The results confirm that the BTS process is capable of producing high purity syngas with adjustable H_2:CO ratio, and that the ITCMO oxygen carrier can effectively crack and convert tars from biomass and the CLHG process is capable of producing high purity H_2 from low purity reducing gas feedstocks. A techno-economic analysis (TEA) focused on a biomass to methanol plant using the BTS process shows an increased process efficiency and decreased methanol required selling price.

Renewable Thermoplastics from Thermochemical Biomass Conversion

Conference: Process Development Symposium

Year: 2020

Proceeding: 2020 Process Development Symposium

Group: General Submissions

Session: Emerging Technologies

Time: Wednesday, June 3, 2020 - 11:25am-11:50am

Authors:

Wilson, N., National Renewable Energy Laboratory

Thermoplastics are pervasive in modern society, and biomass feedstocks offer a sustainable source for these materials. Catalytic fast pyrolysis of biomass is an emerging technology for producing biofuels but accessing high-valued thermoplastic precursors from this process has remained a challenge. To enable this approach, the technical barrier to overcome is the economic isolation of sufficiently pure chemical precursors for materials synthesis. Achieving this goal will require process development that considers the high heterogeneity, increased oxygen content, and unique impurities in biomass conversion streams. Within this work, we will show how well-established separation unit operations can be applied to catalytic fast pyrolysis intermediate streams to access polycarbonate precursors. The impurities in chemical products will differ when using biomass as the feedstock versus using petroleum, and a bioderived catechol will be used as an example to discuss how these changes can impact downstream synthesis. Finally, novel thermoplastics that are synthetically challenging to access via petrochemical routes but are attractive from the catalytic fast pyrolysis of biomass will be presented. This work shows that accessing thermoplastics from thermochemical conversion of biomass can be a viable strategy for creating sustainable materials.

Syngas-to-Olefins: Evaluation of Reaction Kinetics and Optimization of Packed Beds of Mixed Catalysts

Conference: Process Development Symposium

Year: 2020

Proceeding: 2020 Process Development Symposium

Group: General Submissions

Session: Emerging Technologies

Time: Wednesday, June 3, 2020 - 11:50am-12:15pm

Authors:

DeWilde, J., The Dow Chemical Company

Packed beds containing physical mixtures of both Zn-Zr mixed oxide catalyst and SAPO-34 convert syngas directly into a mixture of C2-C5 olefins and paraffins in a single reaction vessel. Specifically, the mixed oxide catalyst is responsible for intermediate oxygenate synthesis from syngas while the molecular sieve catalyzes olefin synthesis from oxygenate intermediates generated on the mixed oxide. Unfortunately for targeted olefin synthesis, olefins are generated in presence of feed hydrogen, causing undesirable direct hydrogenation of the desired olefin product. Kinetic measurements with co-fed propylene over each catalyst, Zn-Zr mixed oxide and SAPO-34, independently confirm olefin hydrogenation activity over both components of the composite bed. While pure hydrogen and olefin feeds demonstrate the mixed oxide catalyst as more strongly catalyzing hydrogenation, the addition of CO to the feed of hydrogen and propylene drops the activity of propylene hydrogenation over the Zn-Zr oxide. Adding water co-feeds to the CO/propylene/H_2 mixture further drops hydrogenation activity. In sum for the target syngas-to-olefins chemistry under reaction conditions of syngas feed and produced water, olefin hydrogenation predominantly occurs over the SAPO-34 catalyst, rather than over the catalyst responsible for hydrogenating CO into oxygenate intermediates. A developed kinetic model consistent with this conclusion describes measurements at differing feed compositions, temperatures, space velocities, and bed catalyst mixing ratios. The developed kinetic model is further extended via optimization to predict the desired catalyst mixing ratios for olefin yield of laboratory scale reactor beds.

Addressing the challenges of a continuous operation at intermediate scale

Conference: Process Development Symposium

Year: 2020

Proceeding: 2020 Process Development Symposium

Group: General Submissions

Session: Case Studies

Time: Wednesday, June 3, 2020 - 1:00pm-1:30pm

Authors:

Ebrahimpour, O., DuPont

Scaling up in multiple stages from a laboratory-developed idea to a commercial-scale process offers several advantages. It mitigates the probability of technology failure at larger scale and it helps to select the right equipment and unit operation options. At the DuPont-Kingston Technology Center (KTC), process development researchers scale up the idea from the lab scale into intermediate scale before the pilot plant or demonstration scale. This is to drive down the cost of a new molecule synthesis at the early stages of the process development and identify and tackle technical risk associated with the proposed technology. Also, at the same time, operating at intermediate scale can reduce the cost of pilot plant operation by optimizing process robustness. However, developing, designing and operating a scalable manufacturing process at intermediate scale is very challenging, particularly for continuous operation. This is due to limited knowledge about the process, safety issues and availability of equipment at this scale. In this presentation, process scale-up approaches and associated challenges at intermediate scale are highlighted. In addition, creative strategies that have been developed successfully in the "Kingston Process Development Group" in DuPont Canada to tackle these challenges are described.

Applications of Fluidized Bed Technology

Conference: Process Development Symposium

Year: 2020

Proceeding: 2020 Process Development Symposium

Group: General Submissions

Session: Case Studies

Time: Wednesday, June 3, 2020 - 1:30pm-2:00pm

Authors:

Freireich, B., Particulate Solid Research, Inc.

Fluidized beds (FB) and circulating fluidized beds (CFB) are widely used in industry. The major processes that utilize FBs and CFBs will be discussed. Each process has evolved over the years with its own design rules and culture. Understanding various processes and how these unit operations are implemented help us understand the best practices and general design guidelines.

This webinar will cover the following topics:

- Particle size, density and Geldart Group classification,
- Unifying Regime phase diagram indicating various regimes of fluidization,
- Fixed bed vs. fluidized beds vs. circulating fluidized beds,
- Catalytic processes,
- Noncatalytic processes,
- Physical processes,
- Regimes of fluidization in each of this process

AUTHOR INDEX

AIChE
120 Wall Street, FL 23
New York, NY 10005-4020

ISBN 978-1-7138-1819-9

CURRAN ASSOCIATES INC.
proceedings.com

Management Division

Held at the 2024 AIChE Annual Meeting

San Diego, California, USA
27-31 October 2024

AIChE
The Global Home of Chemical Engineers